DEPARTMENT OF EDUCATION AND SCIENCE

**An HMI Survey**



# NAFE

***Non-Advanced Further Education***

# in practice

# 8 ENGINEERING



London: Her Majesty's Stationery Office

First published 1988

ISBN 0 11 270645 2

CONTENTS

**Appendices**

## Introduction

1. This report presents the findings of an inspection of non-advanced further education (NAFE) courses in engineering carried out during 1985–86. In 1987, HM Inspectorate published a report[1] on an inspection of NAFE provision in England and Wales. This publication on engineering is one of a series of supplementary reports[2] based on that inspection.

2. Two hundred and forty thousand students attended non-advanced courses in engineering during the academic year of 1985–86; of these 36,000 were full-time, 167,000 part-time and 37,000 evening-only students. Three hundred and ninety-two colleges of further education provide engineering courses in England and Wales. Many of these colleges have more than one engineering department and in 1985 there were 580 engineering departments in further education establishments in England and Wales.

## The inspection sample

3. Thirty-one engineering departments were inspected in the 24 colleges listed in Appendix I. During the inspection 781 classes were visited and over 7,000 students observed. Sixty different courses were inspected. The distribution of students on these courses between the various modes of study was as follows:

| | |
|---|---|
| Full-time and sandwich | 15 per cent |
| Part-time day and evening | 63 per cent |
| Evening only, including short courses for adults | 22 per cent |

---

[1] *NAFE in practice.* HMSO, 1987.

[2] Details of other supplementary reports in the series can be found on page 52 of this publication.

4. Student enrolments in the 31 engineering departments ranged between 203 and 1,873. In 22 of the departments 80 per cent or more of the engineering students were on non-advanced courses. The distribution of classes between the engineering sub-specialisms is shown in Appendix II.

5. In the Summer of 1986, 61 visits were made to employers to obtain their views on the quality of NAFE provision in engineering and of the links which had been established between them and their local colleges.

## *Course provision*

6. The pattern of courses in engineering falls broadly into three categories:

i. vocational education and training for the trainee in employment;

ii. vocational education in full-time college courses for the student who may seek a future career in engineering;

iii. pre-vocational provision through the Certificate of Pre-Vocational Education (CPVE), the Technical and Vocational Education Initiative (TVEI) and linked courses with schools.

In 1985–86, 80,000 students were enrolled on Business & Technician Education Council (BTEC) courses and 157,000 on City and Guilds of London Institute (CGLI) craft courses; the latter figure includes courses for industrial operatives devised by the CGLI and the Engineering Industry Training Board (EITB). Although operatives form the largest group of employees in the engineering industry (Appendix III, Table 1) less than 2 per cent receive any formal education and training. The Youth Training Scheme (YTS) programmes in engineering usually include off-the-job training provided by colleges consisting of the whole or part of CGLI or BTEC courses or pre-vocational courses.

7. Colleges devise courses to suit local needs following guidelines laid down by the CGLI and BTEC. The first year of courses for craft trainees in employment can be full-time if initial training as well as further education takes place in college. Subsequent attendance for craft students is usually part-time and the CGLI Craft Certificate level is normally attained in three years. Successful students may then proceed for a further year's study for the Advanced Craft Certificate which, in turn, can lead to the CGLI licentiateship. It is also possible for the holder of a CGLI Craft Certificate to undertake bridging studies to enter a BTEC technician course. Profiles of typical craft courses are given in Appendix IV.

8. At the time of the inspection, a small group of colleges were offering the first year of a new CGLI pilot craft scheme in Basic Engineering Competencies (CGLI 201) which will replace existing craft schemes over the next few years. This new scheme is intended to provide students with the skills and knowledge needed to satisfy the demands now being made by industry for a flexible and multi-skilled workforce.

9. Courses for technicians lead to BTEC National Certificate or Diploma awards. These courses are modular in design and are made up of interchangeable units of study; the Diploma course entails

the study of a greater number of units than the Certificate course. Both the Certificate and the Diploma course may be studied either full-time or part-time but it is normal practice for the Certificate to be taken on a part-time and the Diploma on a full-time basis, each requiring two years' study. Further study can lead to BTEC Higher National Certificate or Diploma awards (HNC and HND). The National Certificates and Diplomas are also accepted for entry to degree courses in engineering. A profile of a typical technician course is shown in Appendix V.

10. BTEC has recently undertaken a review of its courses and intends to implement certain changes in their structure and organisation. There will be a new pattern of courses leading to four types of qualification: First, National, Higher National and Continuing Education. Entry requirements are to be standardised and limitations will be imposed on modes of study; for example, it will not normally be possible to study full-time for a National Certificate. Certificates of Achievement recognising success in the study of individual units or groups of units are also to be introduced.

11. The EITB and the Road Transport Industry Training Board (RTITB) set standards and specify patterns for training. Trainees in employment are required to undertake industrial training to obtain certification by the Industrial Training Boards (ITB). Usually further education and off-the-job basic training take place in the first year (46 weeks) of an apprenticeship. After the first year, training and practice in selected skills are usually provided in the workplace. Where the first year of apprenticeship training takes place entirely in college it is described as integrated further education and training. At its best this type of provision involves integration of education and training throughout the week. In many colleges, however, the bulk of the further education is

concentrated into one day with training over the other four days. Non-employed students on full-time college courses may also undertake an integrated first year of further education and training but unless they subsequently enter suitable employment in the industry it is not possible for them to complete the full cycle of industrial training required for ITB certification. Nevertheless, such students often progress by part-time studies to CGLI or BTEC awards.

## *Curriculum*

12. There is no single further education curriculum. Rather there are curricular traditions of which that concerned with training and education for industry has most relevance for NAFE provision in engineering. In this tradition curriculum design has been heavily influenced by examining bodies such as the CGLI and the BTEC as well as by the industries themselves through the training boards and the engineering professional bodies.

13. In engineering, the curriculum is more than the formal programme of theoretical, practical and vocational activities associated with a particular course. It is also influenced by such features of the college as the relationships between staff and students; links with local industry and the community; the values exemplified in the way the college is organised and managed; the styles of teaching and learning; and the ethos of a particular department or course. Students learn from all these things, and all should be seen to be consistent with the aims of the college and the course.

14. The practical work undertaken in college as a complement to theoretical studies may take a variety of forms, including

assignments and projects; it is distinct in the curriculum from the training which a student undertakes to acquire competence in industrial skills.

15. The typical day-release craft course involves about four hours each day on craft principles and associated studies, one or two hours of general studies and two or three hours of practical work. In a day-release technician course it is usual for five BTEC units to be studied during the day at college. The balance between theoretical and practical work usually depends on the approach adopted by the teacher. The full-time BTEC courses vary in pattern but normally a student studies about 10 BTEC units during a week over about 20 hours, undertakes simulated work experience for five hours and spends between five and 10 hours on practical training in the college workshops.

16. Although CGLI craft courses contain general studies units, no examinations are taken in this subject and colleges are free to devise their own syllabuses. CGLI courses in communication skills (CGLI 772) and numeracy (CGLI 364) are often used by colleges as a form of general studies.

17. Work experience is an important feature of some full-time engineering education. Ideally, this takes place with an engineering employer but, owing to a shortage of such training places, many colleges and training centres provide simulated work experience. Students derive particular benefit from work experience which is based on a structured programme of work covering several facets of a company's activities. Work experience is valued by both students and teachers as a means of providing a realistic context in which theoretical studies may be put to the test; in addition, work

experience helps students to gain a greater understanding of the engineering industry and furthers their personal and social development.

18. In some full-time BTEC courses, training elements had been incorporated which satisfied in part requirements for simulated work experience but fell short of those of the ITBs. Other BTEC work experience units involved programmes of industrial visits or short periods (two to four weeks) of work experience with local firms. Although students found such experience valuable it was more advantageous to them when it could be credited as training acceptable to the ITB.

19. The syllabuses of CGLI and BTEC courses are specified through long lists of objectives and are designed by working groups of teachers and industrialists. Teachers may develop individual teaching schemes but these seldom include topics dealing with broader competencies or an understanding of issues beyond the boundaries of engineering.

20. Few engineering departments had established procedures for curriculum review and evaluation. Discussion of curricular issues, when it took place, was usually informal. Curriculum review was seldom carried out by course teams at the direction of course tutors. In one large engineering department, however, a strong course team had made a particular effort to monitor its full-time technician courses and ensure they were relevant to their students' needs. Most of the students aimed to enter employment rather than higher education, and, as a result, the department offered BTEC Certificate rather than Diploma courses. The smaller number of units in the Certificate course enabled the college to offer a substantial amount of workshop training and this had been

accredited by the ITB. In this college, particular attention was given to ensuring a relationship between practical work in the laboratory and workshop and related theory.

21. In some colleges curriculum evaluation and development stemmed in part from the help and guidance from large local employers in making engineering courses industrially relevant. Meetings of industrial sponsors and teachers were held to discuss course content and teaching methodology. Changes to the curriculum were made by joint agreement between the college and the firms sponsoring students. Evaluation was concerned primarily with the technical content of courses and the sequence of its presentation.

22. There were some deficiencies in the interpretation and development of technical aspects of the curriculum. In general, colleges failed to incorporate new technologies into courses sufficiently quickly; in addition teachers seldom worked together to identify common themes in different subjects.

23. In electronics courses, an understanding of information technology (IT) and its associated equipment is a fundamental requirement of the curriculum; in electronic laboratories the use of microprocessors, including links with robots and other peripheral devices, is common. In other engineering subjects, however, the impact of IT, both on teaching and learning approaches and on course content, was disappointingly low. Few colleges provided opportunities for students to use computers in subjects such as instrumentation and control, general laboratory work, drawing and automobile servicing. In some colleges there was a shortage of

microcomputers and teachers lacked the knowledge, skills and experience to enable them to introduce new technologies within their courses.

24. The introduction of the CVPE, two-year YTS and TVEI had stimulated some successful and valuable curriculum development. For example, one YTS general engineering course had been carefully designed to incorporate relevant aspects of CGLI schemes in numeracy, communication skills and the technology of skilled processes.

## *The engineering industry*

25. Employment in the engineering industry is varied in its nature and scope. Its workforce includes not only specialist engineers but also others who service the wider requirements of the industry. In addition other manufacturing industries (e.g., textiles, chemicals, food and agriculture) also employ engineers as do local authorities and area health authorities. Although the number of people employed by the engineering industry has declined by about one-third since 1978, the industry still accounts for a substantial proportion of the country's labour force. EITB statistics on employment are given in Appendix III. Table 1 of Appendix III shows that in 1986 just under 2,000,000 people were employed by firms linked to the EITB and of these over 600,000 were working as engineering craftsmen, technicians and supervisors. There is a similar number of skilled engineers with firms outside the scope of the EITB; these include employees in shipbuilding, road transport and vehicle maintenance, electrical contracting, electronics and television and radio servicing. Utilities and corporations such as British Telecom, the electricity generation and supply industry, iron and steel, the National Coal Board and the gas and water industries all employ engineers in large numbers.

## *Students*

26. Most engineering students in the sample colleges lived within 10 miles of the institution. However, some colleges had much larger catchment areas, either because they were relatively isolated or because they served as regional centres for certain courses.

27. In several colleges there were many students from ethnic minority groups on full-time courses but relatively few on day-release courses.

28. Unemployed students can attend vocational courses under the '21-hour' rule of the Department of Health and Social Security (DHSS), by which arrangement they register as unemployed and draw benefit providing they do not attend college for more than 21 hours each week. In one college in the North West of England, 13 per cent of engineering students attended under this arrangement. Students on '21-hour rule' courses who left as jobs became available did not take the final examination. One college, however, offered the '21-hour rule' students a modular course which allowed them to attempt certification at various times during the year.

29. A substantial number of YTS trainees attend the early stages of craft and technician courses. In some cases these trainees joined existing classes under an arrangement known as 'infilling'; in others separate classes were provided for YTS trainees. The motor vehicle industry, for example, uses the YTS programme as the main entry route for its trainees and colleges usually offered special classes for motor vehicle engineering trainees.

30. A few colleges recruit a substantial number of overseas students. For example, in full-time craft and technician courses in refrigeration and air conditioning at a large North London college,

overseas students accounted for nearly 50 per cent of the enrolments. These students attended courses on a full-cost basis and their commitment to their course of study was impressive.

31. Students usually had appropriate entry qualifications for their course of study and some departments took great care in matching students' initial qualifications and aptitude to an appropriate course of study. The selection procedure often included interviews with prospective students and their parents, and some colleges required students to complete a diagnostic test.

32. In most instances, the selection procedures of employers helped to ensure that part-time students were appropriately qualified for the courses they followed. However, employers sometimes recruit craft apprentices with substantially better qualifications than those required for entry to the course. Such apprentices sometimes find craft courses insufficiently challenging.

33. By contrast, some colleges accepted under-qualified students onto full-time technician courses such as the BTEC National Diploma. Many of these students had difficulty in meeting the demands of their course and either failed to complete it or were awarded a Certificate instead of the Diploma. Where there were substantial numbers of such students in a course, the overall standards achieved were often low.

34. The majority of NAFE students in engineering departments were between 16 and 22 years of age. Of the older students attending courses, some were unemployed and attended under the DHSS '21-hour rule'. When they formed part of a mixed age group, these older students often brought the benefits of maturity

to classroom discussions and other course activities. Older students also attended evening courses, including those providing updating in specialist knowledge and skills.

35. Engineering students are almost exclusively male. However, a few colleges have attempted to redress the balance by developing courses for female students; for example, the EITB-sponsored 'women in engineering' short course is designed to provide guidance on engineering as a subject and as a career.

36. Most students were well informed about their courses and gave sensible reasons for choice of course and career. Amongst the day-release groups, the craft students found their course relevant to their employment. By contrast, the technician students generally believed that most of their studies would be relevant at a later stage in their careers.

37. Many full-time students on BTEC courses hoped to progress to higher education at universities and polytechnics. Over 80 per cent of students on one such course had firm intentions to proceed to higher education. Some full-time students believed higher education qualifications would enhance their chances of obtaining appropriate employment. Part-time students on release from industry usually intended to continue in their present employment after completing their course and a number expressed an interest in continuing to higher-level craft or technician studies. Most full-time BTEC students had been made aware by their teachers that higher education at university or polytechnic was an option; this was, however, seldom the case with part-time students although having obtained their National Certificates many of the more able

were eligible for degree or HND studies. Unemployed students and YTS trainees usually believed that their course led to a qualification and provided experience which would help them to obtain a job.

38. Most students expressed satisfaction with their course of study and with the support received from college staff. Although some colleges had a flourishing programme of students' union activities, engineering students rarely participated in them; to some extent this was because engineering students were heavily timetabled and a high proportion attended on a part-time basis.

## *Careers education and counselling*

39. In general, engineering students had made a vocational choice before choosing their course. Part-time students in employment normally had determined their own career plans. The nature of the guidance offered to full-time students depended on whether they wished to continue in full-time education or intended to seek employment immediately after the completion of their course. Full-time students on BTEC courses who wished to progress to higher education were advised on the combination of units required for entry to a Higher National Diploma or degree course; sometimes these students were advised to undertake extra studies in mathematics, to improve their prospects of entry to and success in higher education. The course tutor had usually established an information bank on higher education establishments and was able to provide details of their requirements. He also served as the focus of help and advice for full-time students seeking employment. Class teachers provided another source of careers advice often acquired through their links with local industry and with former students who maintained contact with the college after entering employment. Full-time students also gained valuable knowledge about the engineering industry through work experience

placements. In spite of the wide range of advice available to them, however, some students had unrealistic ambitions or were unclear about future options open to them.

40. The majority of departments held open days when local employers and prospective students visited the college. The open days provided employers with an opportunity to see students' work and to assess its quality. A number of departments invited industrialists to talk to full-time students about career opportunities open to them. It was also common practice for senior college staff to visit schools to talk about careers in engineering.

## *Teaching and learning*

41. In most of the colleges the majority of lessons were well-prepared and the general standard of classroom practice was satisfactory. In four colleges, however, a substantial number of lessons were poorly taught. Although much of the teaching was sound and conscientious, it was often unimaginative; students were required to copy notes, either from the blackboard or through the sterile process of dictation, and they were fed information they could have better discovered for themselves. In the best practice, teachers combined a sound knowledge of their subject with a lively teaching style which engaged and held the students' interest; lessons were conducted with a sense of purpose and students were afforded a variety of appropriate learning activities. Practical work usually commanded the students' attention and enthusiasm because it required their full participation and enabled them to work at their own pace.

42. In most courses students were timetabled for long working days, and for day-release students in particular a succession of technical theory lessons taxed their concentration and patience. The

students accepted the pattern and style of teaching to which they had become accustomed and relationships between teachers and students were usually good.

43. Many of the BTEC day-release courses contained insufficient practical work; often this was because teachers had reacted to the demands of full syllabuses by concentrating on classroom teaching. There were few examples of teaching which took account of and drew upon the work experience of technician students. To some extent this was because the teachers themselves lacked recent industrial experience; many former craft teachers now teaching on technician courses had no direct industrial experience as technicians.

*Classroom and practical work*

44. Workshop and laboratory sessions were generally lively and purposeful. In addition to participating in specific practical exercises or projects the students engaged in discussion with their teachers and sought their advice when necessary. By contrast, classroom sessions were often dominated by teachers whose prime aim was to transmit information which the students diligently copied into their notebooks.

45. Carefully designed handouts were used effectively in some motor vehicle and fabrication and welding classes; these handouts elicited response from the students and made them think carefully about information given. In some instances, they formed a valuable supplement to, or substitute for, notes.

46. In general, students were not sufficiently involved in classroom sessions. In one instance, the photocopied section of a descriptive text was distributed and then read through by the

teacher in its entirety; the students lost interest. On another occasion, a part-time technician class watched patiently while the teacher toiled through calculations on the chalkboard which, with benefit to the students, they could have worked out for themselves.

47. In the better classroom practice, assignments were used not only for assessment purposes, but also as a means for students to discover information for themselves. The assignment placed responsibility on the student for learning a particular part of the syllabus; it was also used to encourage students to develop and express their own views, to marshal their own evidence and to draw their own conclusions. Students worked on assignments individually or in groups. For example, in a communications studies assignment a group of craft students undertook the selection and presentation of written information, formed a discussion group and sustained a lively and disciplined debate for an hour.

48. Students also benefited from project work. Most projects involved teamwork and, particularly where they required some form of construction or assembly, took days or even weeks to complete. Projects were effective in reinforcing learning and allowing students to exercise and develop a range of skills but in some colleges they were not carefully designed. Teachers were sometimes unable to explain the aims and objectives of the projects and some tasks stood little chance of being completed in the time available. However, some excellent work took place that was properly structured. For example, in a one-year microelectronics evening course for mature students in employment, after introductory lectures on the theory of computer architecture, each student was required to design and build a microcomputer from basic components and then to test it. Assembly involved wiring and soldering the various sub-systems, after which peripherals, such as push-buttons and memory circuitry, were added as desired.

Inevitably each computer had different characteristics. In this project students progressed at their own pace in a task which required delicate manual skills and a fundamental understanding of systems theory. The success of this lesson rested in part on the ability of the teacher to respond accurately and positively to a variety of questions. By contrast, in another project class in which microprocessor open-learning kits were being used by students to write simple programs, the teacher arrived late and ill-prepared; he was not able to deal adequately with students' questions and there was a general air of confusion throughout the class.

49. In most colleges, students enjoyed laboratory work which engaged their full participation. The more effective laboratory sessions illustrated specific theoretical principles which had been the subject of recent classwork. For example, a two-hour lesson on electromagnetic induction theory was divided into three equal sessions. After the teacher had introduced the theory, the class moved to the laboratory where a series of illustrative experiments had been set up ranging from simple magnets to a car ignition coil opened up to show windings and provided with a spark gap. The teacher demonstrated the equipment briefly, after which the students carried out experiments. The teacher dealt with queries effectively and encouraged discussion on topics which arose during their practical work. A final classroom analysis of the experimental findings helped to consolidate the students' understanding of this part of their syllabus.

50. On occasions, it proved as effective for students to observe a practical demonstration as to carry out the work themselves. For instance, in one lesson students learnt much from, and displayed evident interest in, a well-presented demonstration of the process of tempering a cold chisel; in another lesson, students similarly benefited from a demonstration of the use of a voltage regulator.

By contrast, in a lesson on the air–fuel flow in engines, the teacher failed to make use of some simple demonstration models available in the room and gave the students a rather confused and ineffective blackboard presentation instead.

*Tutorials and private study*

51. Few courses, either part-time or full-time, had timetabled tutorial periods during which teachers provided individual students with academic guidance or counselling. Most teachers were ready during normal lessons to provide their students with whatever individual help and advice they needed, but most were unenthusiastic about timetabled tutorial periods.

52. Few departments had policies on the setting of homework. Individual teachers decided how often to set homework; some classes were set none and few were given more than one homework exercise every three or four weeks. Homework usually consisted of exercises to consolidate what had been taught in theory classes.

53. Engineering students seldom made regular use of college libraries. Part-time students had little free time for library work, which teachers seldom encouraged. For full-time students, a well planned assignment or project sometimes involved a literature search, but in most colleges there were too few library staff to enable consistent professional library support to be provided in specialist fields.

## Assessment and standards

54. Students were keen to obtain engineering qualifications. The majority of advertised engineering jobs require particular levels of qualifications and experience. Once in a job, promotion prospects and increased responsibilities are often dependent on continued academic progress.

55. Examinations are set by the CGLI at the end of the first and third years of the craft courses (Part I and Part II) and at the end of the advanced craft courses (Part III). The Regional Examination Bodies (REB) also set Part I examinations. Colleges are required to assess coursework. The Part I examination takes the form of a multi-choice question paper; the answers do not require more than single sentences, phrases or words and students do not find the paper difficult. At this level the onus is on the teachers to set good standards for coursework. The Part II and Part III examinations require longer written answers to questions and provide a greater challenge to the students.

56. The BTEC provides colleges with guidelines on assessment and requires details of assessment to be included by colleges in the course schemes submitted for validation; colleges determine their own assessment schemes. Each subject unit within a programme is usually assessed through coursework, assignments and periodic tests (phase tests) and an end-of-course examination. All marking is carried out by teachers. External moderators appointed by BTEC must approve the assessments carried out by a college. Assessment schemes observed during the survey contained an appropriate number of phase tests (usually two). The phase tests provided a valuable early indication of a student's progress and some teachers had used them diagnostically to identify students' needs for appropriate remedial teaching.

57. There was considerable emphasis on practical applications in all courses. CGLI schemes required students to undertake practical coursework and assignments and such activities featured in the sample colleges. By contrast there was a low incidence of practical and laboratory work in BTEC courses; although reference was made to laboratory work in assessment schemes, these mainly concentrated on phase tests and examinations.

58. There was some variation between engineering departments in the standards of assessment applied in BTEC courses. Moderators appointed by BTEC were able to eliminate some of the more obvious differences; for example, by pointing out where examination papers varied in order of difficulty and where projects and assignments differed not only in the complexity of task set but in the marking. Few engineering departments had established effective monitoring of their assessment procedures.

59. The CGLI emphasises the importance of practical coursework but does not require it to be externally assessed. The standards of practical work in CGLI courses varied considerably from college to college. Although most departments had course teams, these seldom addressed themselves to the identification and monitoring of standards expected of students in their practical work.

60. In the small amount of CPVE and TVEI work seen, no assessment was observed. The engineering teachers were expected to contribute to student profile assessments on these courses and they recognised their need for in-service training in these new assessment processes.

61. There were large variations between colleges in the success rates of students and in the methods used to record them. Some colleges, in calculating overall success rates, recorded examination

passes but ignored general wastage during the course. Some colleges measured success in BTEC courses through the number of passes in individual units rather than the number of students who had achieved a BTEC award.

62. The practice of recording pass records for each BTEC unit allows identification of units which students find difficult. For example, in one full-time BTEC Diploma course the failure rates in electrical principles and in mathematics units were consistently 20 per cent higher than in other units. In one college a 70 per cent failure rate in CGLI Electrical Engineering Craft Studies was attributed to the fact that a large block of students had been required to work overtime by their employer. The college had brought this matter to the attention of the employer concerned.

63. Some colleges were achieving high student retention rates and good examination results. For example, one college claimed only a 0.5 per cent drop-out on BTEC courses followed by an average pass of 90 per cent in individual units. In another college, however, of 37 students who enrolled on a BTEC Diploma in 1983 only 21 took examinations in 1985 and of those only five were awarded a Diploma. High failure levels of this type were due in part to the practice of enrolling students with minimal entry qualifications. However, there were examples amongst the sample colleges in which notable success had been achieved with such an entry.

64. Appendix VI shows cohort analyses normalised to a first-year intake of 100 students for CGLI and BTEC students. The success rate in minimum course time was 53 per cent for Part II CGLI craft students, 49 per cent for BTEC Diploma students and 65 per cent and 34 per cent for two-year and three-year part-time BTEC Certificate students respectively. The main causes of wastage are

withdrawal from the course during the academic year because of redundancy, general employment uncertainties, the rejection by students of engineering as a career during early stages of their courses and examination failure. Some students transfer to other courses. In addition, some trainees remain in engineering employment but do not continue their education. Examples of best practice indicate that efficient monitoring of academic progress and counselling of students in difficulty can significantly improve success rates.

## *Organisation and management*

65. In most of the colleges inspected, engineering was the largest specialist area of work. Some colleges based all engineering in a single department and others had separate departments for each specialism (for example, mechanical, electrical and motor vehicle engineering). Departments were normally organised on a conventional line management structure, with the head of department working through specialist section leaders. Most departments had senior teachers in charge of craft and technician sections and in the larger departments it was usual for groups of teachers in a particular craft or subject area to be led by a senior teacher.

66. All of the departments were adjusting to rapid changes; these fall into three categories:

i. the overall decline in student numbers;

ii. the effect of new technology on industrial practice and consequently on technical aspects of the curriculum;

iii. new national educational initiatives, particularly those requiring an introduction to engineering in pre-vocational courses.

In some departments, changes in staffing, management structure and accommodation have enabled them to react flexibly and appropriately to changes, but in others adjustments to traditional provision have been unacceptably slow.

### *Attendance and timetabling*

67. Full-time students usually attend college for up to 30 hours a week and part-time day students attend for up to 11 hours. Part-time day attendance usually starts at 9.00am and finishes at about 8.30pm or 9.00pm. Part-time day BTEC students often have to attend on an additional evening because of the number or combination of units which they require. In the departments inspected, all engineering students were heavily programmed and there was little time available for private study. The part-time evening-only courses were usually programmed for two or three evenings a week, each class lasting for about two-and-a-half hours.

68. Engineering courses are usually timetabled in lessons of between one-and-a-half and two hours. Such a length of period is suitable for practical work, but it is often too long for classroom sessions. The claim by teachers that they are short of time to complete the syllabuses stems in part from the inflexible and inefficient timetabling.

### *Group sizes*

69. Appendix VII shows the distribution of the class sizes observed during the survey. Classes varied from groups of two or three students to groups of over 20. Some of the larger groups were divided for practical work. *In several colleges, between half and two-thirds of the classes had fewer than 10 students.* There were some instances where related classes were combined in order to achieve larger teaching groups. For example, some full-time and part-time

BTEC groups were combined to achieve viable numbers for classes dealing with the same units. Unfortunately there was seldom any attempt to exploit the different backgrounds and experience of the students in such mixed classes, and teachers missed the opportunity to draw upon the industrial experience of part-time students for the benefit of full-time students.

70. Some of the training classes observed were too large to be effectively and safely supervised by one teacher. For example, in one college YTS students in groups of 15 or more were using welding equipment and a variety of machine tools. As a result the students did not receive the constant supervision which they needed at this stage in their training.

## Resources

### *Staffing*

71. Departments were adequately staffed with well-qualified and experienced teachers although some specialist areas such as electronic engineering were under strength. By contrast *in areas such as mechanical and production engineering there was over-staffing.* Most teachers had industrial experience but the majority had been teaching for 10 or more years and their industrial experience was dated. Approximately half of the teachers in the sample colleges were over 50 years of age. Because of reductions in staff complements in recent years, there were relatively few part-time teachers; where such teachers were employed their knowledge of current industrial practice compensated in part for the out-dated experience of many full-time teachers.

72. Technicians provide an essential support service for the specialist activities of engineering departments. They undertake the maintenance and repair of equipment in workshops and laboratories, ordering of materials and the construction of specialist

items of equipment for students in specialist laboratories. Often, engineering technicians also had responsibility in their own departments, and more widely in the college, for the maintenance and repair of electronic equipment such as computers and audio-visual aids. The technicians were fitted for that work by experience and many also had technician and craft qualifications. The proportion of technician to academic staff varied from 0.66 to 0.15 with an average of 0.33. Technician staff establishments in engineering tend to be more generous than in most subject areas because of the complexity of engineering equipment and the need to maintain and service large laboratories and workshops. Where the level of technician support was good, teachers were able to provide demonstrations of equipment and materials more frequently in the classroom and these helped the students to carry out their practical work more effectively. Engineering departments often found difficulty in recruiting and retaining technicians in new technology areas such as microelectronics and computing applications.

73. The level of administrative support was generally insufficient, so that teachers had to carry out routine clerical and time-consuming administrative tasks such as filing, students' records and administrative duties associated with enrolment. Most departments had one secretary for the head of department, with no additional clerical support for teachers. One head of department was using a computer to store information. He was able to provide statistics quickly on such matters as student records and teaching workloads.

### *Staff development*

74. Few of the departments inspected had a formal procedure for identifying or meeting the in-service needs of either teaching or support staff. Nevertheless, most colleges recognise that staff development is essential if teachers are to respond to changing demands. Staff development in the sample colleges was largely a

result of individual initiatives. This took a variety of forms, including attendance at both specialist engineering and teacher training courses; visits to industry; work with professional bodies; work as specialist examiners and moderators; and service on both local authority, regional examining body and professional committees. The recent Department of Trade and Industry (DTI) grants for the purchase of computer numerically controlled (CNC) equipment in colleges had required staff to undertake training courses in the use of this equipment and this initiative had stimulated both staff and curriculum development.

75. Efforts to release teachers for long in-service training using Department of Education and Science (DES) In-service Training (INSET) pooling arrangements were severely hampered by difficulties experienced in obtaining suitable temporary replacement staff; it was particularly difficult to obtain temporary staff in those subjects where retraining of existing staff is needed, such as electronics, computer-aided engineering and software engineering. Although staff in craft subjects were sometimes under-utilised, few were undertaking retraining in shortage subjects.

## *Accommodation and equipment*

76. Although most departments were accommodated in purpose-built post-war buildings, these usually suited the educational demands of engineering in the 1960s rather than the present. For example, many engineering workshops still contain large machine tool workshops equipped with lathes, milling machines and other production equipment formerly used by large craft apprentice intakes; some workshops also contain heat engines and electrical machinery laboratories which are seldom used. The need for such accommodation and resources continues to fall as industry adapts to the lower manpower needs of high technology production lines. Few colleges have carried out systematic reviews of their range of

equipment in order to identify obsolete items which might be discarded, thereby freeing valuable space. The trend in engineering education is towards teaching through practical work and learning through hands-on practical experience with modern technology. Provision of funds through central government schemes such as the DTI machine tools initiative has enabled colleges to make a start on re-equipping the workshops with CNC tools. Some colleges use this equipment to provide mid-career training for employees of local industry.

77. Falling student numbers in engineering and growth in other areas of college provision have required engineering departments to release classrooms formerly designated solely for engineering for wider college use; as a result, the engineers are increasingly having to use workshops and laboratories as general teaching areas. Although this sometimes has the advantage of helping the teacher to relate theory and practical work, the practical areas often have poor seating arrangements which are not conducive to effective learning. For example, theory lessons were seen in laboratories which had benches and tall backless stools arranged around the perimeter of the room. These were uncomfortable and made note-taking difficult and students could not see the chalkboard easily. These problems were often compounded when laboratory equipment restricted use of the bench surfaces. In some departments, practical areas had been redesigned and refurbished to accommodate both practical and classroom activities: sensible seating as well as equipment benches and materials display areas were provided. Such accommodation was particularly suitable for demonstrations by teachers and for project and assignment work.

78. Many classrooms were drab, poorly furnished and in need of decoration. Even in classrooms dedicated to engineering there were

seldom any relevant engineering wall displays. Most classrooms were equipped with overhead projectors and these were frequently used by teachers.

79. Few colleges had a rolling programme for the replacement of capital equipment and there was often inadequate provision in college budgets for the maintenance of equipment. Many of the large machine tools were out of date and, although still serviceable, were not representative of modern industrial practice. Departments which have obtained new equipment have often done so using income generated by full-cost courses or as a result of direct-funding initiatives by the Government. In one department, 70 per cent of funding came from sources other than the local authority and there is an increasing tendency for departments to rely on such external funds.

### *Libraries*

80. The quality of library provision varied widely between colleges. Some engineering teachers encouraged library searches in project and assignment work but this often strained library resources. Engineering textbooks in about one-third of the libraries visited were out of date. There was some evidence that librarians received insufficient help from specialist teachers in keeping the book stocks up to date.

## *External relations and marketing*

81. Most colleges had made considerable efforts to establish and maintain appropriate links with local employers. Representatives from local engineering employers are invited to sit on departmental advisory committees where they have the opportunity to comment on the structure and quality of existing course provision and to advise on student recruitment strategies and industry's future needs.

82. Links with industry have been developed primarily through part-time day-release education and the nature of local industry has had a considerable effect in determining the scope of educational provision in engineering departments. Where one major company provides a large number of students, the courses are usually tailored to the needs of that company; company training programmes are often matched to, and compatible with, those provided by the college and the company training officers sometimes take a close interest in curricular matters. Many of the companies involved in this type of association with the colleges are from the public sector, such as British Coal, British Rail and British Steel. The reduction in recruitment by the public sector industries has meant that engineering departments associated with them have faced a substantial decline in enrolments which, in turn, has contributed to the under-utilisation of resources in areas such as machine tool workshops.

83. Some departments provided educational services for a large number of relatively small employers with a wide range of activities. However, industrial involvement with the college tends to be slight in these circumstances because many of the smaller employers do not employ a training officer; in addition, it is difficult for the college to make specific provision for each employer because of their diversity of interests.

84. As part of the survey, visits were made to 61 engineering employers in companies of varying size to obtain their views on

NAFE provision. Those colleges which were held in high regard by employers provided concise information about student attainment and teachers made regular visits to the firms and companies. Many employers would have liked more detailed reports of student performance. The general view of employers was that college provision was good value for money and they considered the college staff they dealt with to be competent, reliable and co-operative. Some employers expressed concern over the difficulties that colleges experience in keeping their equipment up to date and the apparent inflexibility of college budgeting systems.

85. In recent years the MSC has been funding provision in further education through initiatives such as the Youth Opportunities and Training Opportunities Programmes, and latterly through the YTS. The MSC has required engineering departments to relate courses more directly to the training requirements of industry. Although engineering departments have always had close links with local industry, their collaboration with the MSC has sharpened their perception of industrial needs other than those covered by traditional training and further education.

86. The Engineering Council's policy statement *Standards and Routes to Registration*, published in 1984, defined the requirements for the registration of technicians. However, most NAFE students have little contact with the professional bodies and relatively few of them seek registration after they have qualified. Academic staff are usually members of the appropriate professional body and some individuals play a leading role in the local branch organisations.

87. Recent initiatives in pre-vocational education, such as the CVPE and TVEI, have strengthened the collaboration between schools and further education colleges. Another consequence of pre-vocational initiatives has been the general revival of school link

courses in some colleges. These new client groups have brought other advantages. For example, where craft teachers have a central role in the new provisions this has helped offset the under-utilisation of such teachers resulting from the fall in traditional craft student recruitment. In addition, the development of teaching methods has been stimulated by the need to cater for students with requirements which differ from those of the traditional engineering student.

88. Apart from links with local industry and schools, few departments have developed significant links with the local community. In one college, however, some engineering teachers had helped in the provision of education facilities for the mentally or physically handicapped and in another a college had developed a short full-time course under the local Community Programme which dealt with the refurbishment of domestic electric appliances. The appliances were subsequently offered to the local Social Services department for use in rehousing schemes for the homeless. Students on this course were drawn from local prisons and psychiatric hospitals and the course helped in their rehabilitation. The teachers had made a considerable effort to devise an appropriate curriculum and had visited the local Community Programme workshop in the course of this work.

89. Some engineering departments have found it difficult to come to terms with the decline in demand for certain courses and the need to pursue a more vigorous marketing policy. For example, colleges have done relatively little to inform the schools, young people and their parents of the scope and flexibility of full-time technician courses. These provide good preparation for employment and an entry qualification for higher education. A number of colleges offer programmes of short courses to local engineering

employees; there is, however, scope for more provision of this type and for consultation with local industry to determine how its needs can be met through short courses.

## *Conclusions*

90. Engineering is not only one of the largest but is also one of the longest established areas of provision in NAFE. It has been subject to considerable change and, in recent years, this has accelerated because of the impact of IT upon the nature and scope of employment in the industry. Few colleges are adapting sufficiently quickly to this change. In some, equipment, course content and staff expertise have become out of date and classroom practice is insufficiently imaginative or appropriate. Some colleges, which are responding conscientiously and responsibly, are establishing new courses and curricula.

91. The bulk of the work in college engineering departments has traditionally been concerned with part-time courses for apprentices. Changes in the needs and recruitment patterns of the industry have, however, been reflected in a decline in demand for some well-established but narrowly specialised courses. Students now require a wide range of skills and the ability to be adaptable in order to be effective members of a flexible, multi-skilled engineering workforce. In particular, all aspects of engineering are increasingly concerned with electronics and the application of computers. Colleges are aware of these trends which are reflected in some effective course development but, in general, there is insufficient emphasis on new and changing technologies within the curriculum. Although information technology is a central feature of electronics courses, it has not penetrated the curriculum of other engineering specialisms to any significant extent.

92. Engineering teachers are usually suited, by qualifications and experience, to teach on part-time courses that are job-specific in their nature and scope. However, there is increasing demand from students who wish to undertake full-time studies before entering employment or continuing in further education. There has been some innovative and successful curriculum development in this field, but few colleges give sufficient priority to the designing of successful full-time courses.

93. The majority of students on the courses inspected were employed males between 16 and 22 years of age, and they attended college on a part-time basis. However, there was a significant number of unemployed students on many part-time courses. These unemployed students, as well as YTS trainees, saw the acquisition of an engineering qualification as a passport to a job. Many students on full-time technician courses intended to proceed to higher education but this aspiration often resulted from low immediate employment prospects. Students were generally, but not always, appropriately qualified. Although most students received appropriate pre-enrolment guidance and were placed on courses which suited their needs, the abilities and ambitions of a number of craft trainees would have been better matched on a craft rather than the technician course which they had joined. Some students were uncertain about career and educational opportunities open to them and they would have benefited from more systematic counselling and guidance. Student morale was, however, generally good and relationships between students and teachers were excellent.

94. Classroom teaching was usually effective in providing a good grounding in engineering but was often unimaginative in its presentation. Many teachers had not properly developed the basic syllabus guidelines of BTEC and CGLI into structured schemes of work. Few departments had a clear policy on curriculum evaluation

and course tutors were seldom able to talk with conviction about course aims. Many engineering courses have over-full syllabuses which most teachers are anxious to cover adequately. As a result, there was seldom enough time for the exploration of broader issues relating to the nature and scope of engineering or for the exercise of skills such as communication and problem-solving, which further the students' personal and social development. When invited to join discussions or participate in group work, however, students reponded with enthusiasm. Practical work was often the best feature of courses; it invariably sustained the interest of students and some of them were carrying out work of a high standard. Work experience is valuable for full-time students; unfortunately, although several colleges had been notably successful in placing their students with suitable firms, work experience has become increasingly difficult to arrange. Valuable simulated work experience and training to ITB specifications were being carried out in many of the colleges. Most engineering students were heavily timetabled, and there were insufficient opportunities for tutorials and private work; there was little regular use made of libraries. There were under 10 students in approximately one-third of the classes seem. However, where full-time and part-time classes were combined in order to form a cost-effective group, there was seldom an attempt by teachers to adapt their teaching styles to ensure that the needs of all students were met.

95. Projects and assignments are increasingly becoming a good feature of engineering courses. Properly organised, they place responsibility on the students for the management of their own learning and for developing and exercising skills which are necessary in industry. Some excellent work was seen, but some teachers fail to set clear and feasible aims and objectives in these exercises.

96. Continuous assessment in BTEC courses by teachers provides them, and the students, with an indication of progress achieved; it also serves to identify weaknesses which need to be remedied, and this function is much appreciated by the students. There is, however, scope for clearer definition and control of national standards to ensure that these are recognised and understood by teachers making assessments locally. The CGLI sets and marks examinations externally but this excludes the teachers from a significant part of the assessment. Student success in achieving qualifications varied considerably between colleges but, measured over minimum length, many courses had unacceptably high wastage rates of 50 per cent or more. Many students achieved qualifications by extending their studies beyond the normal length of the course.

97. Teachers were generally well qualified but their industrial experience was often dated. There was a surplus of teachers with traditional manufacturing experience but a widespread shortage of electronic engineers. In both non-advanced and advanced further education, the shortage of electronics teachers is now affecting other subject areas which are increasingly dependent on IT. Few engineering departments had staff development policies for the updating of teachers' first-hand knowledge of new technologies and modern industry; such staff development as occurred was usually the result of initiatives taken by individuals. It was difficult for departments to find adequate replacements for teachers released to undertake in-service training. Because of the decline in recruitment to engineering courses in recent years, and the consequent reduction in staff complements, engineering departments employ few part-time teaching staff; however, where part-time staff were employed, they brought a knowledge of current industrial practice which partially compensated for the outdated experience of many full-time teachers. Technician

support in engineering was often insufficient to meet all the demands made upon it; these include the preparation of classrooms, laboratories and workshops, support for teaching and demonstrations and the servicing and maintenance of equipment, often on a college-wide basis and including some of a highly sophisticated kind. Too often teachers were required to undertake technician and simple clerical duties.

98. Most of the departments inspected in this survey are accommodated in purpose-built post-war buildings. However, the nature and design of the accommodation are usually more suitable for the engineering courses provided in the 1960s than for the new high technology engineering of today. A few departments are still housed in converted pre-war annexes sometimes remote from the main college buildings. Many colleges still retain largely outdated and unused machine workshops, electrical machines and heat engine laboratories. Recent injections of specific funding such as the DTI CNC machine tool initiative have enabled many colleges to purchase specialist equipment. In some colleges the local education authority has provided funds for the purchase of the equipment necessary to develop computer-aided design/computer-aided manufacture (CAD/CAM) and robotics courses. Much of the equipment in NAFE engineering departments is, however, obsolete and does not fully meet the needs of students who work in the modern engineering industry. Few colleges were able to carry out a systematic review of their range of equipment or plan a replacement programme with confidence that the necessary funding would be available.

99. Engineering teachers maintain good contacts with local industry and many also serve as moderators and examiners for the BTEC and CGLI. Links with the MSC are established largely through the negotiation of full-cost courses and YTS provision.

New initiatives such as the CPVE and TVEI are resulting in closer collaboration between schools and colleges. Many employers believe that engineering courses are good value for money. Few engineering departments, however, are making enough effort to market their courses in order to ensure that schools, young people and their parents know and understand the range of NAFE provision available to those who seek employment or who wish to enter higher education.

# Appendix I

## The colleges inspected

Brooklyn Technical College
Chippenham Technical College
Crawley College of Technology
Doncaster Institute of Higher Education
Erith College of Technology
Gateshead Technical College
Great Yarmouth College of Further Education
Guildford College of Technology
Hackney College
Handsworth Technical College
Kilburn Polytechnic
Lancaster and Morecambe College of Further Education
Neath College
New College, Durham
North Nottinghamshire College of Further Education
Peterborough College of Adult Education
Plymouth College of Adult Education
Rotherham College of Arts and Technology
St Helens College
South Cheshire College
Tameside College of Technology
Tile Hill College of Further Education
Willesden College of Technology
Ystrad Mynach College of Further Education

## *Appendix II* Distribution of classes between engineering sub-specialisms in the inspection sample

Mechanical: 32 per cent A
Electrical: 26 per cent B
Motor Vehicle: 25 per cent C
Fabrication and Welding: 5 per cent D
General Engineering: 7 per cent E
Marine, Aeronautical, Pattern Making and Foundry: 5 per cent F

BTEC courses: 43 per cent of work seen
CGLI courses: 41 per cent of work seen

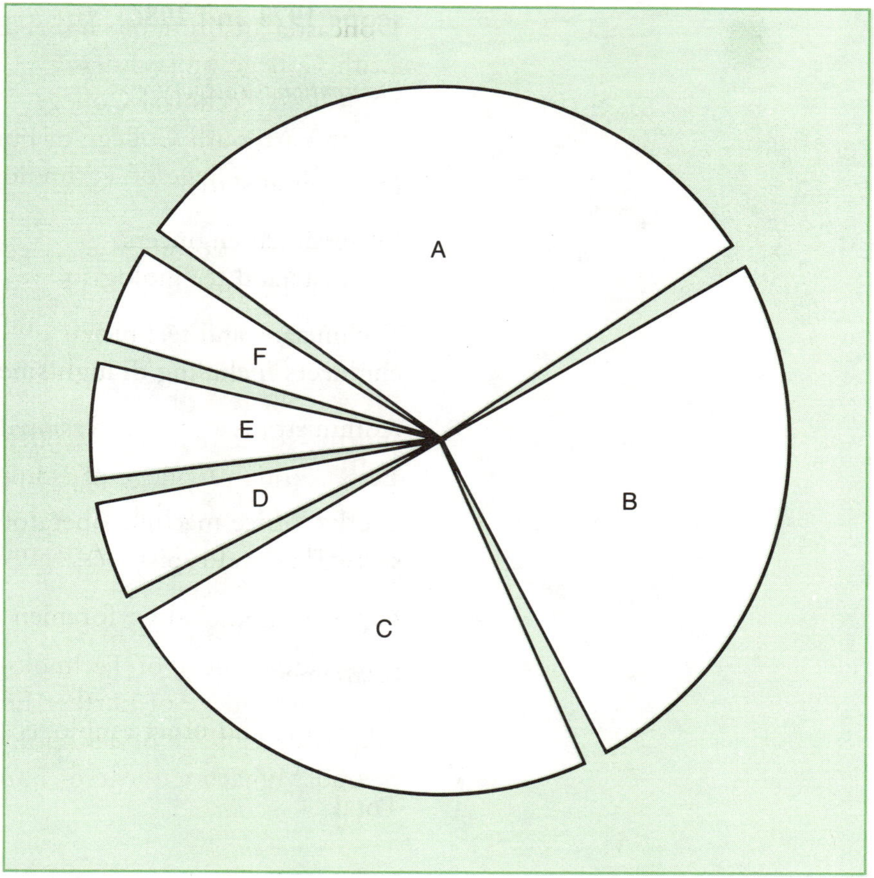

# Appendix III

## Engineering employment statistics

### Table 1

**Employment by occupation in engineering firms in EITB scope 1978 and 1986**

| *Occupational category* | *Employees (thousands)* *1978* | *1986* | *% change* *1978 to 1985* |
|---|---|---|---|
| Managerial staff | 133 | 122 | – 8 |
| Professional engineers, scientists and technologists | 58 | 89 | + 52 |
| Technicians and technician engineers including draughtsmen | 212 | 181 | – 15 |
| Administrative and professional staff | 147 | 134 | – 9 |
| Clerks, office machine operators, secretaries and typists | 333 | 197 | – 41 |
| Supervisors including foremen | 145 | 94 | – 35 |
| Craftsmen | 532 | 342 | – 36 |
| Operators and other employees | 1,379 | 799 | – 42 |
| Total | 2,939 | 1,956 | – 33 |

**Table 2**

**Employment by sector in engineering firms in EITB scope 1986**

| *Sector* | *Employees (thousands)* |
|---|---|
| Metal manufacture | 52 |
| Foundries | 59 |
| Other metal goods | 220 |
| Industrial plant and steelwork | 57 |
| Machine tools | 56 |
| Machinery | 432 |
| Office and data processing equipment | 83 |
| Basic electrical equipment | 89 |
| Electronics | 247 |
| Other electrical engineering | 151 |
| Motor vehicles | 239 |
| Other vehicles | 34 |
| Aerospace | 159 |
| Instrument engineering | 64 |
| Other activities in scope[1] | 14 |
| Total | 1,956 |

[1] 'Other activities in scope' refer largely to specialised research establishments and research consultancies.

**Table 3**

**Employment by region in engineering firms in EITB scope 1986**

| *Region* | *Employees (thousands)* |
|---|---|
| South East | 595 |
| East Anglia | 58 |
| South West | 150 |
| West Midlands | 357 |
| East Midlands | 144 |
| Yorkshire & Humberside | 135 |
| North West | 235 |
| Northern | 77 |
| Wales | 75 |
| Scotland | 130 |
| Total | 1,956 |

**Table 4**

**Employment by size of engineering establishment in engineering firms in EITB scope 1986**

| *Size of establishment (number of employees)* | *Number employed (thousands)* | *Number of establishments (thousands)* |
|---|---|---|
| 1–24 | 128 | 11.2 |
| 25–40 | 116 | 3.6 |
| 41–99 | 210 | 3.2 |
| 100–249 | 319 | 2.0 |
| 250–499 | 296 | 0.9 |
| 500–999 | 278 | 0.4 |
| 1,000–4,999 | 473 | 0.3 |
| 5,000 + | 136 | < 0.1 |
| Total | 1,956 | 21.7 |

*Source:* EITB statutory returns; 1986 figures are provisional.

# *Appendix IV*

## Full-time Craft Level Course

## Examples of engineering craft courses

### *Motor Vehicle Craft Studies CGLI 381 Part I*

(Similar courses exist for other specialisms and for general engineering.)

| | |
|---|---|
| *Target students:* | Those completing secondary education. |
| *Duration:* | One year (46 weeks). |
| *Entry requirements:* | Motivation to become employed in retail or fleet motor vehicle servicing. |
| *Mode of attendance:* | Full-time. |
| *Awarding body:* | CGLI and/or college certificate. |
| *Content:* | Craft Principles, Associated Studies and Activities and General Studies (8 – 12 hours/week). Practical Skill Training (12 – 18 hours/week) (often based on RTITB scheme). Other pre-vocational studies and activities (0 – 6 hours/week). |
| *Assessment:* | Continuous assessment of practical skills to RTITB standards. End-of-course examination by CGLI or REB. |
| *Certificate:* | CGLI (Part I) certificate is sometimes issued for the course as a whole. |
| *Progression:* | To employment and by further study to CGLI Motor Vehicle Craft Studies Part II. |

*Course developments:* At the time of the inspection, the YTS scheme was absorbing increasing numbers of those who would otherwise have found employment directly in the industry or who would have attended full-time courses such as described above. The industry is increasingly using YTS as its normal source of recruits and two-year YTS is likely to confirm this trend. The current revision of CGLI Craft Studies courses has not yet affected the CGLI 381 series.

### *CGLI 205 Part II Craft Certificate in Mechanical Engineering*

*Target group:* Mechanical and Production Engineering Craft Trainees who have completed CGLI 200 Basic Engineering Craft Studies. While the course is designed for such trainees, who will normally be undertaking module training under EITB arrangements, it also accommodates other trainees who may be receiving less well-planned training and these may form a substantial proportion of some groups.

*Duration:* Part II is normally a two-year part-time release course. It therefore occupies the second and third years of training. It consists notionally of about 480 hours of technical studies and 120 to 180 hours of general studies.

*Entry requirements:* Successful completion of CGLI 200.

*Mode of attendance:* Mainly part-time day and block-release.

*Examining body:* CGLI.

*Content:* The Part II course builds on the broad base of Part I with Craft Principles, Associated Studies, Related Practical Activities and Project Work in the following specialist topics: dimensional control; workholding; machine tools; materials; material removal; planning; general studies; and safety. The scheme intention is that each topic should be taught in an integrated way with as much student involvement as possible.

*Assessment:* Continuous assessment of theoretical and practical coursework. End-of-course examination set by the CGLI.

*Certification:* CGLI (Part II) Craft Certificate.

*Progression:* CGLI (Part III) Craft Certificate or further Part II work or BTEC Certificate or Diploma.

*Course developments:* Following the Hargreaves Report, the CGLI 200 course has been reconstructed to accommodate a wider range of clientele who use engineering craft skills to varying extents whether they are in employment with planned training, in YTS or in other pre-vocational education and training. These new first-year courses are now called Basic Engineering Competencies (CGLI 201). In 1986–87 the first pilot schemes at Part II level were being operated.

# *Appendix V*

## BTEC National Certificate in Engineering with electronics engineering specialism

### An example of a technician course

*Target students:* Trainee technicians in electronics engineering or others intending to become technicians in electronics engineering.

*Duration:* Normally 2–3 years depending on mode of study and academic entry qualifications.

*Entry requirements:* Undertaking technician training in industry (or intending to do so) and relevant GCE–CSE awards of CSE III or better. Former CGLI students may also be admitted with exemption from some units according to previous achievements.

*Mode of attendance:* Part-time release for employed trainees; full-time for unemployed intending technicians. Other modes, e.g. as part of a YTS scheme or within the CPVE, are possible especially for the early stages of the course.

*Awarding body:* BTEC.

*Content:* Typically 15 units (each of 60 hours' design length) at three levels, some of which may be covered by exemption due to previous study. There may be options depending on the college scheme. The unit titles include: physical science level I; mathematics levels I–III; workshop processes and materials level I; general and communication studies levels I–II; electronics and microelectronics systems and principles including digital techniques levels I–III;

reliability, measurements and testing level III. In the full-time course there is a requirement for practical work in lieu of industrial experience. This typically occupies up to four half-days each week.

*Assessment:* Assessments consist of some or all of phase tests, end examinations and assignments, all subject to approval by BTEC-appointed visiting moderators.

*Certification:* The National Certificate is awarded by BTEC to those students who satisfactorily complete 15 units.

*Progression:* For some students the National Certificate marks completion of the FE relevant to their employment. Progress to Higher National Awards (Certificate or Diploma) is usual for others. A minority of students who have high-level passes in certain units proceed to degree courses.

*Developments:* At the time of the inspection, most BTEC courses were operating under validation arrangements made before the specialist Programme Committees of BTEC were abolished and most engineering courses were being revised under the single generic title 'Engineering'. This, and the trend to core studies, together with a reconstruction of the pattern of BTEC awards (Circular 15), heralds a period of change and innovation.

# *Appendix VI*

## Examples of student progression and withdrawal

**Figure 1**

**Student progression and withdrawal in CGLI Engineering Craft course up to Part II normalised to an initial intake of 100**

1. Part I (one year)

Enrolment of 100

Withdrawn 12
Failed examination 10

Awarded Part I Certificate
78

Withdrawn after Part I 9

2. Part II (two years)

Enrolment of 69

Withdrawn 7
Failed examination 9

Awarded Part II Certificate
53

**Figure 2**

**Student progression and withdrawal in BTEC National Certificate and Diploma courses normalised to initial intakes of 100**

1. National Certificate, part-time, two years

Enrolment of 100

| | |
|---|---|
| Withdrawn or failed examinations | 35 |

Awarded Certificate
65

2. National Certificate, part-time, three years

Enrolment of 100

| | |
|---|---|
| Withdrawn or failed examinations | 66 |

Awarded Certificate
34

3. National Diploma, full-time, two years

Enrolment of 100

| | |
|---|---|
| Withdrawn | 19 |
| Transfer | 5 |
| Failed examinations | 27 |

Awarded Diploma
49

*Note:* some students in categories 1 and 2 withdrew and transferred to other engineering courses, but the statistics supplied by colleges gave insufficient detail to determine the numbers involved.

# *Appendix VII*

## Class size distribution in the inspection sample

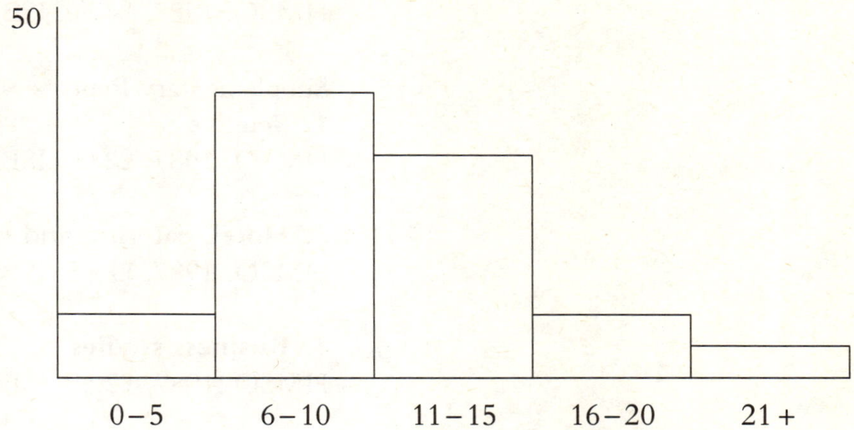


Range of class sizes
i.e. number of students in attendance

## HMI Survey Reports on NAFE

**NAFE in practice**
HMSO. 1987. £4.50. ISBN 0 11 270609 6

Supplementary Reports

**1. Science**
HMSO. 1987. £2.00. ISBN 0 11 270610 X

**2. Hotel, catering and bakery**
HMSO. 1987. £2.95. ISBN 0 11 270623 1

**3. Business studies**
HMSO. 1987. £2.50. ISBN 0 11 270625 8

**4. Agriculture**
HMSO. 1987. £2.50. ISBN 0 11 270634 7

**5. Child, health and social care courses**
HMSO. 1987. £2.00. ISBN 0 11 270636 3

**6. Computing and mathematics**
HMSO. 1987. £2.50. ISBN 0 11 270639 8

**7. Art and design**
HMSO. 1988. £2.50. ISBN 0 11 270640 1

Further details of these and other DES priced publications can be obtained from P9D, Room 2C03, HMSO, St Crispins, Duke Street, Norwich NR3 1PD.

Printed in the United Kingdom for Her Majesty's Stationery Office
Dd. 289295 C32 2/88 405 5673